This Maths book belongs to

_____

Maths book
Grid 10x14 Dark
Book Size 7.5 x 9.25 in
Includes Times tables 1-12

This image cannot currently be displayed.

www.ingramcontent.com/pod-product-compliance
Lightning Source LLC
Chambersburg PA
CBHW080229180526
45158CB00008BA/2255